MW01317510

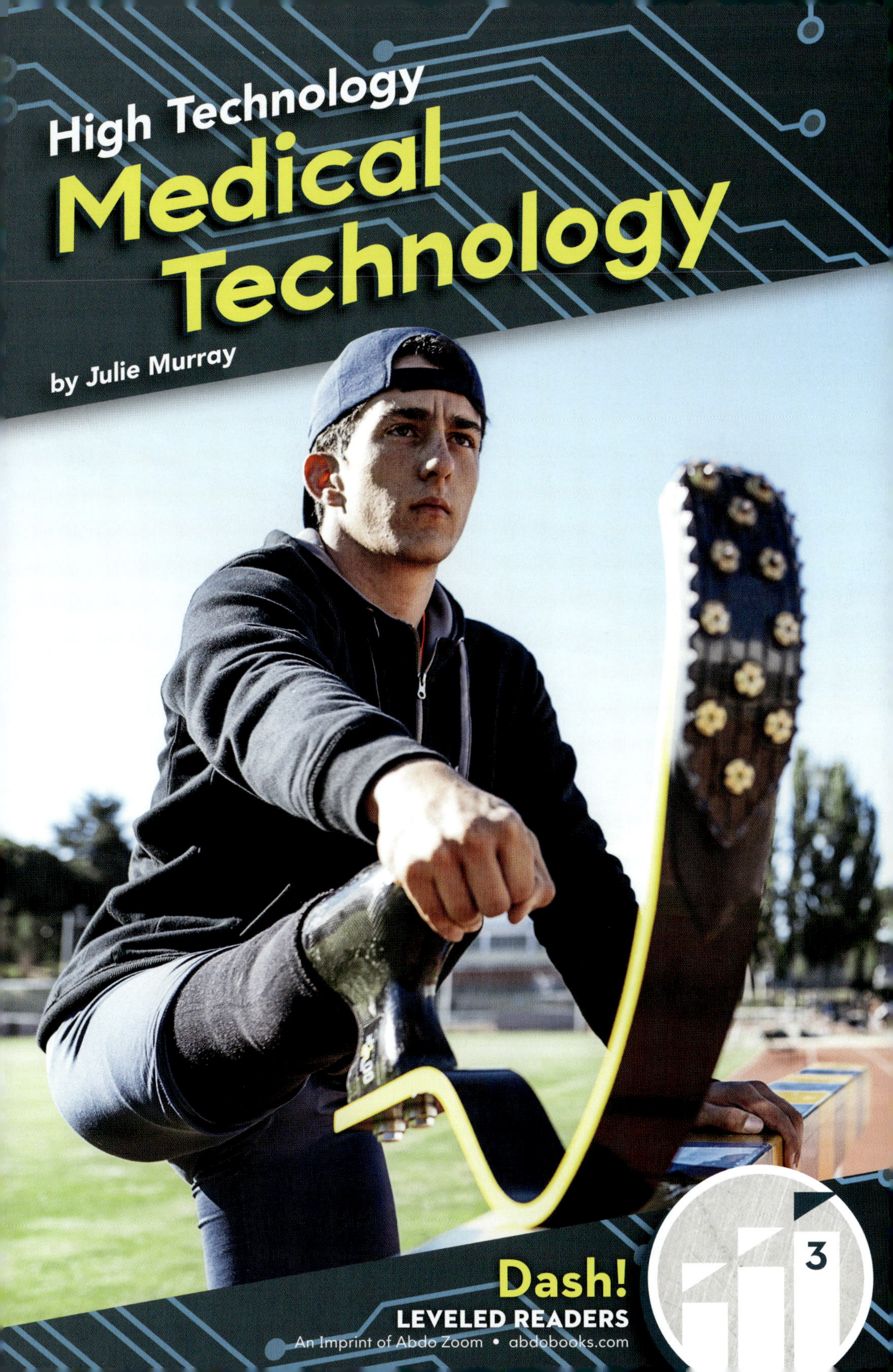
High Technology
Medical Technology
by Julie Murray
3
Dash!
LEVELED READERS
An Imprint of Abdo Zoom • abdobooks.com

Level 1 – Beginning
Short and simple sentences with familiar words or patterns for children who are beginning to understand how letters and sounds go together.

Level 2 – Emerging
Longer words and sentences with more complex language patterns for readers who are practicing common words and letter sounds.

Level 3 – Transitional
More developed language and vocabulary for readers who are becoming more independent.

abdobooks.com

Published by Abdo Zoom, a division of ABDO, PO Box 398166, Minneapolis, Minnesota 55439.

Printed in the United States of America, North Mankato, Minnesota.
052020
092020

Photo Credits: Alamy, Getty Images, iStock, Shutterstock
Production Contributors: Kenny Abdo, Jennie Forsberg, Grace Hansen, John Hansen
Design Contributors: Dorothy Toth, Neil Klinepier, Laura Graphenteen

Library of Congress Control Number: 2019956190

Publisher's Cataloging in Publication Data

Names: Murray, Julie, author.
Title: Medical technology / by Julie Murray
Description: Minneapolis, Minnesota : Abdo Zoom, 2021 | Series: High technology | Includes online resources and index.
Identifiers: ISBN 9781098221171 (lib. bdg.) | ISBN 9781098222154 (ebook) | ISBN 9781098222642 (Read-to-Me ebook)
Subjects: LCSH: Medical technology--Juvenile literature. | Health care technology--Juvenile literature. | Medical innovations--Juvenile literature. | High technology--Juvenile literature. | Technological innovations--Juvenile literature.
Classification: DDC 610.284--dc23

Table of Contents

Medical Technology

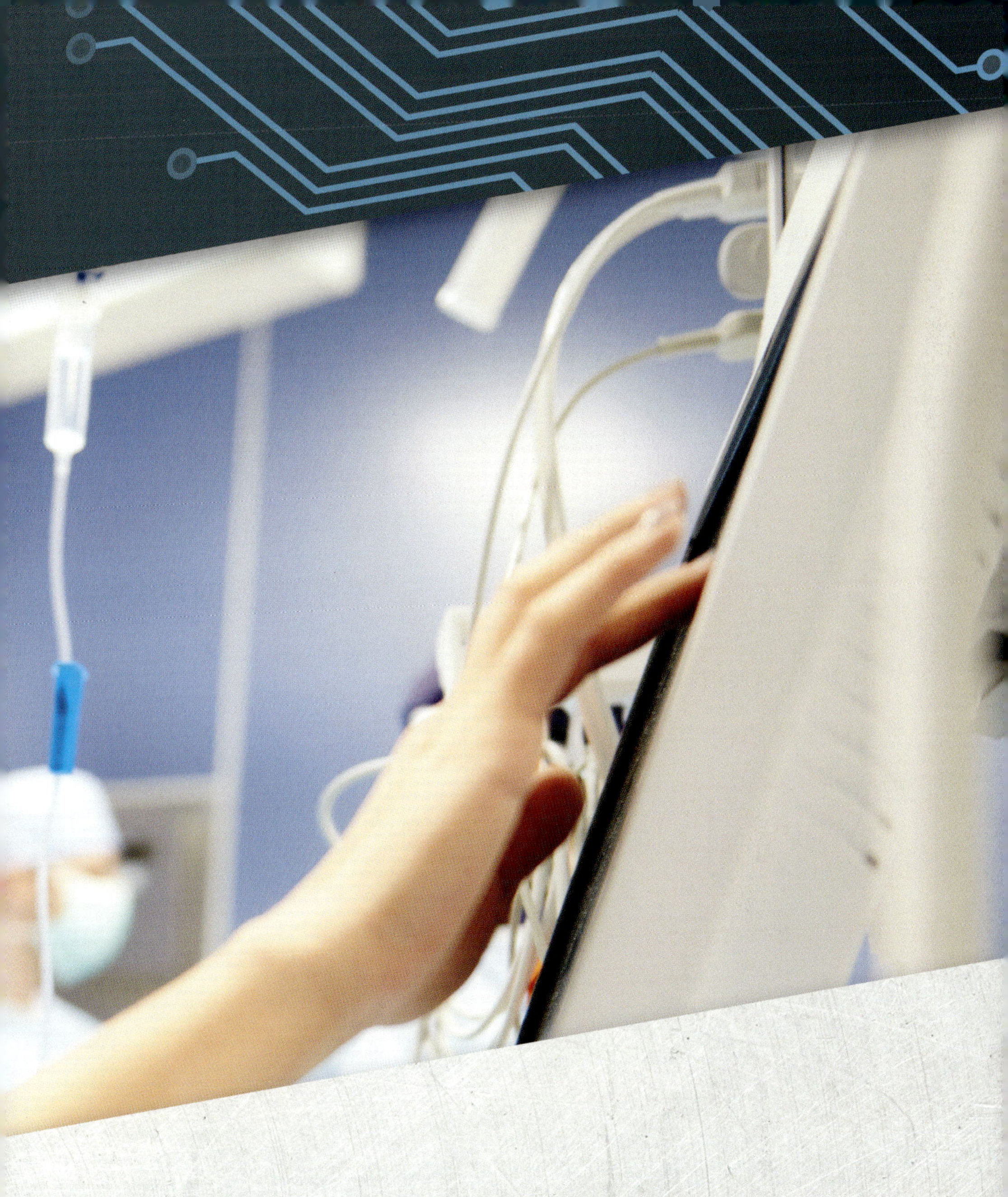

Medical technology is used to save and improve people's lives. It allows doctors to better **diagnose** and treat their patients.

Medical technology includes everything from Band-Aids and eye glasses to heart pumps and artificial limbs.

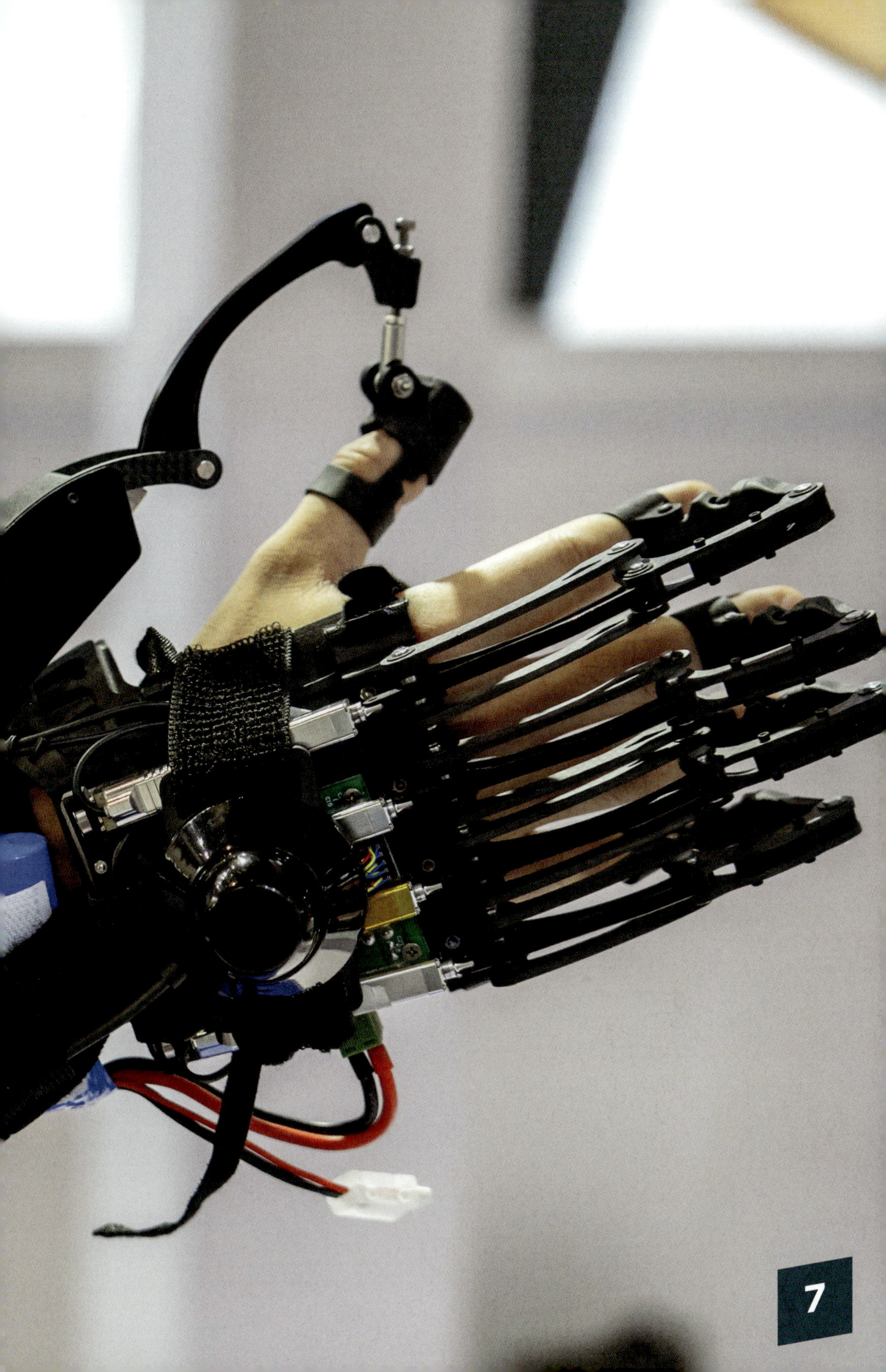

Everyday Uses

Medical technology can be found throughout your doctor's office. A thermometer is used to measure your temperature. A **stethoscope** is used to listen to your heart and lungs.

Medical technology is all around! An **x-ray** machine can see inside your body. Medicines can be taken to make you feel better.

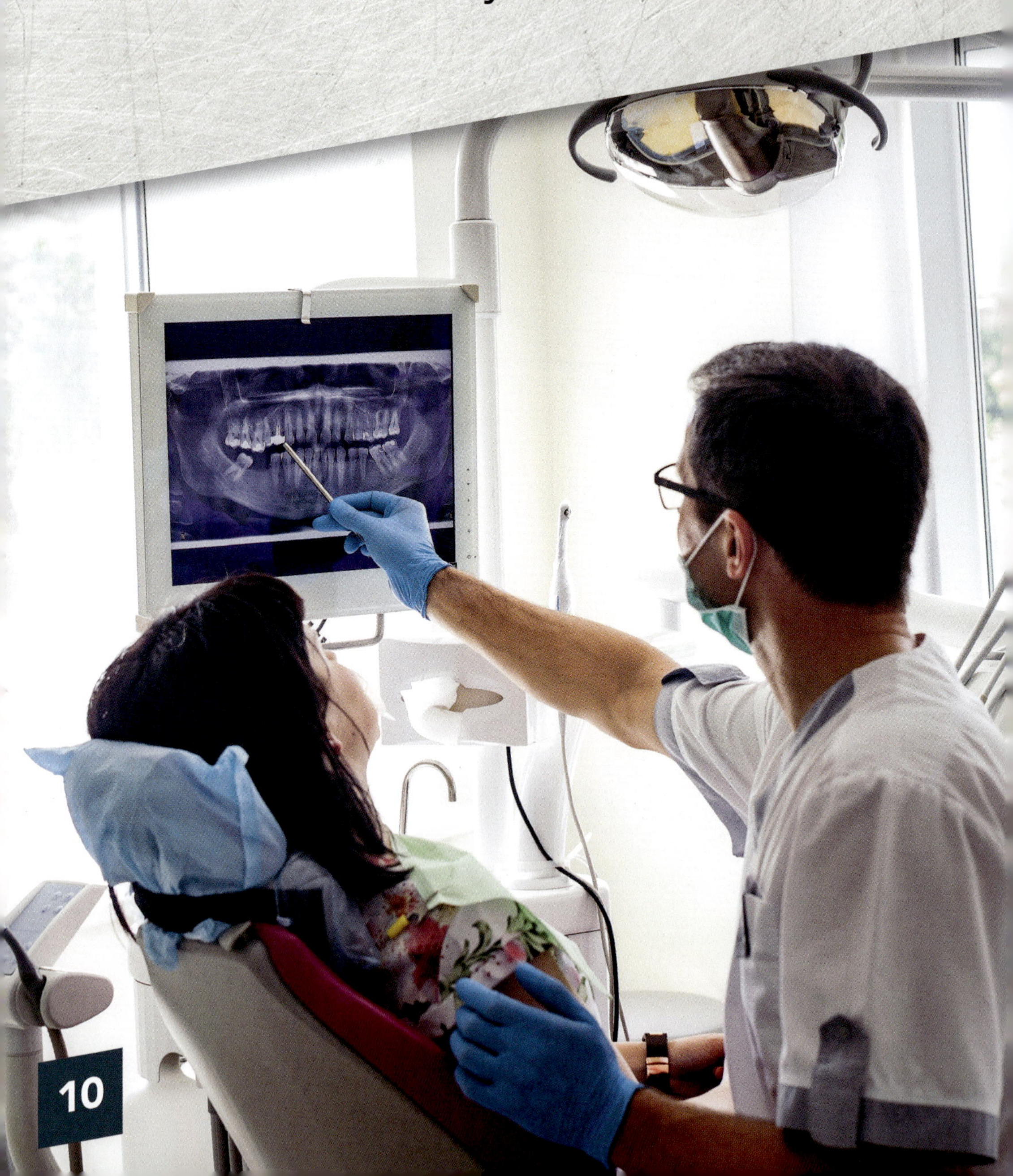

What Can They Do?

Magnetic resonance imaging (MRI) shows tissues inside the body. It can look at organs and the brain. It can even spot **cancer**.

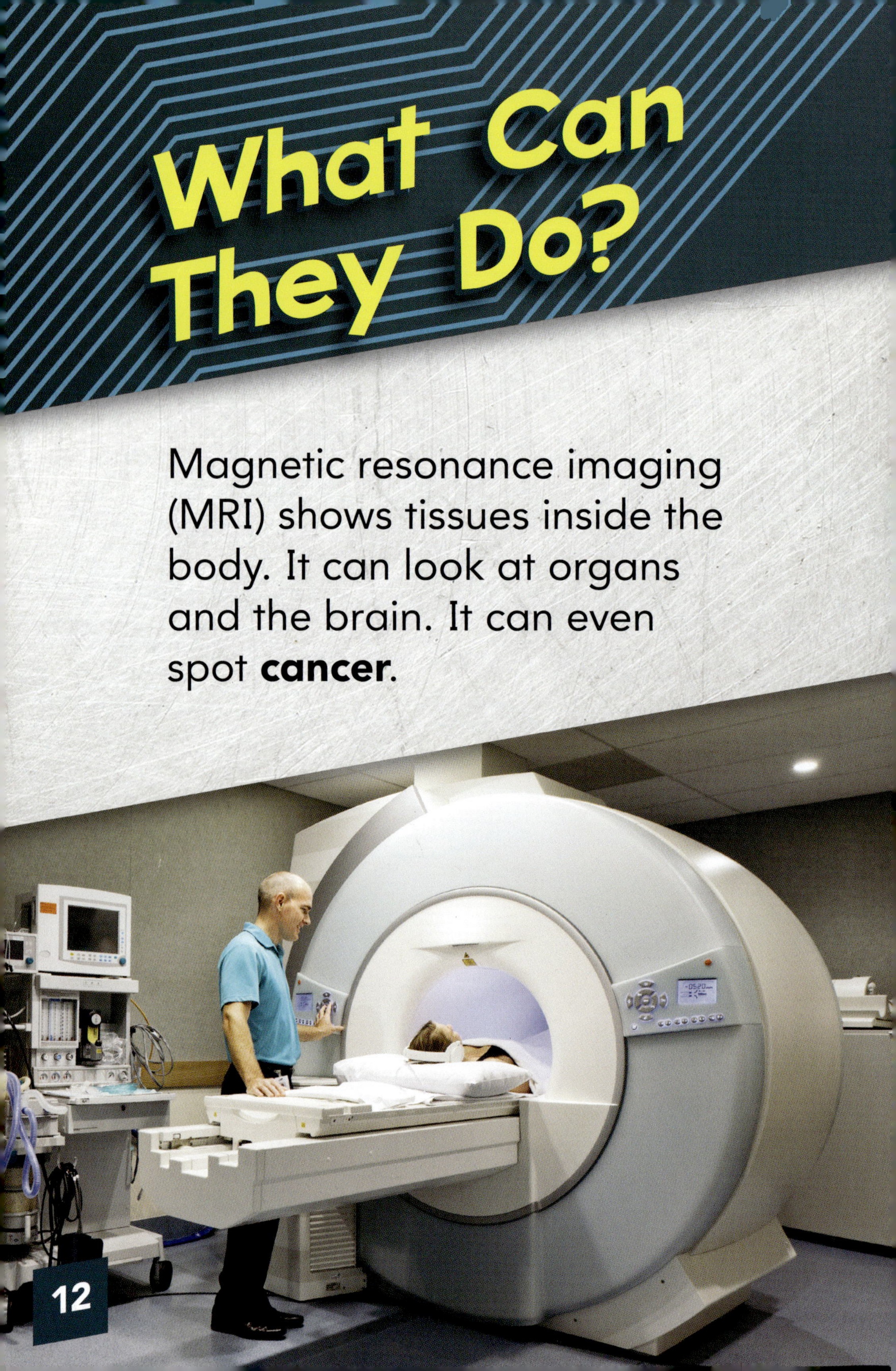

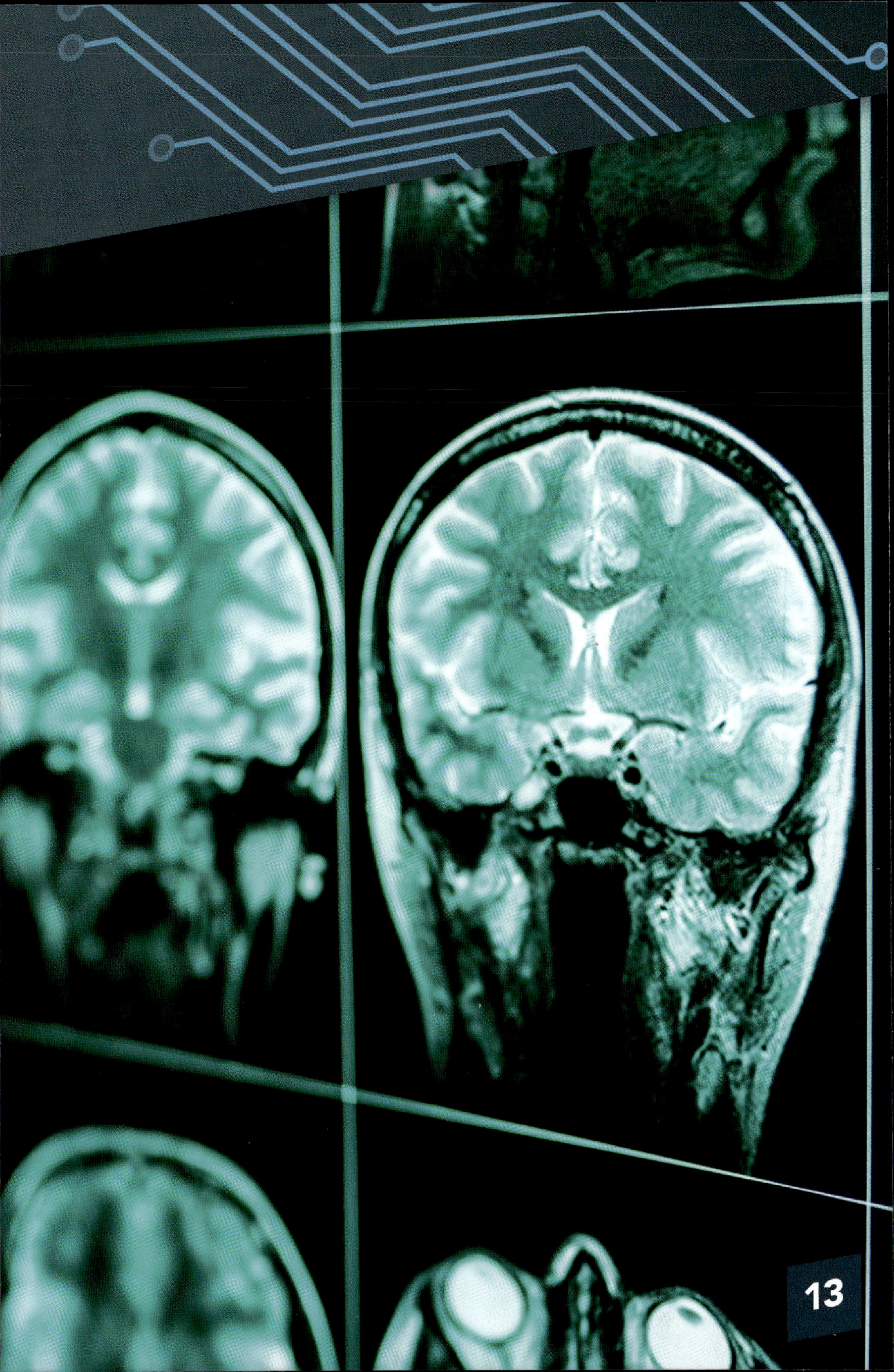

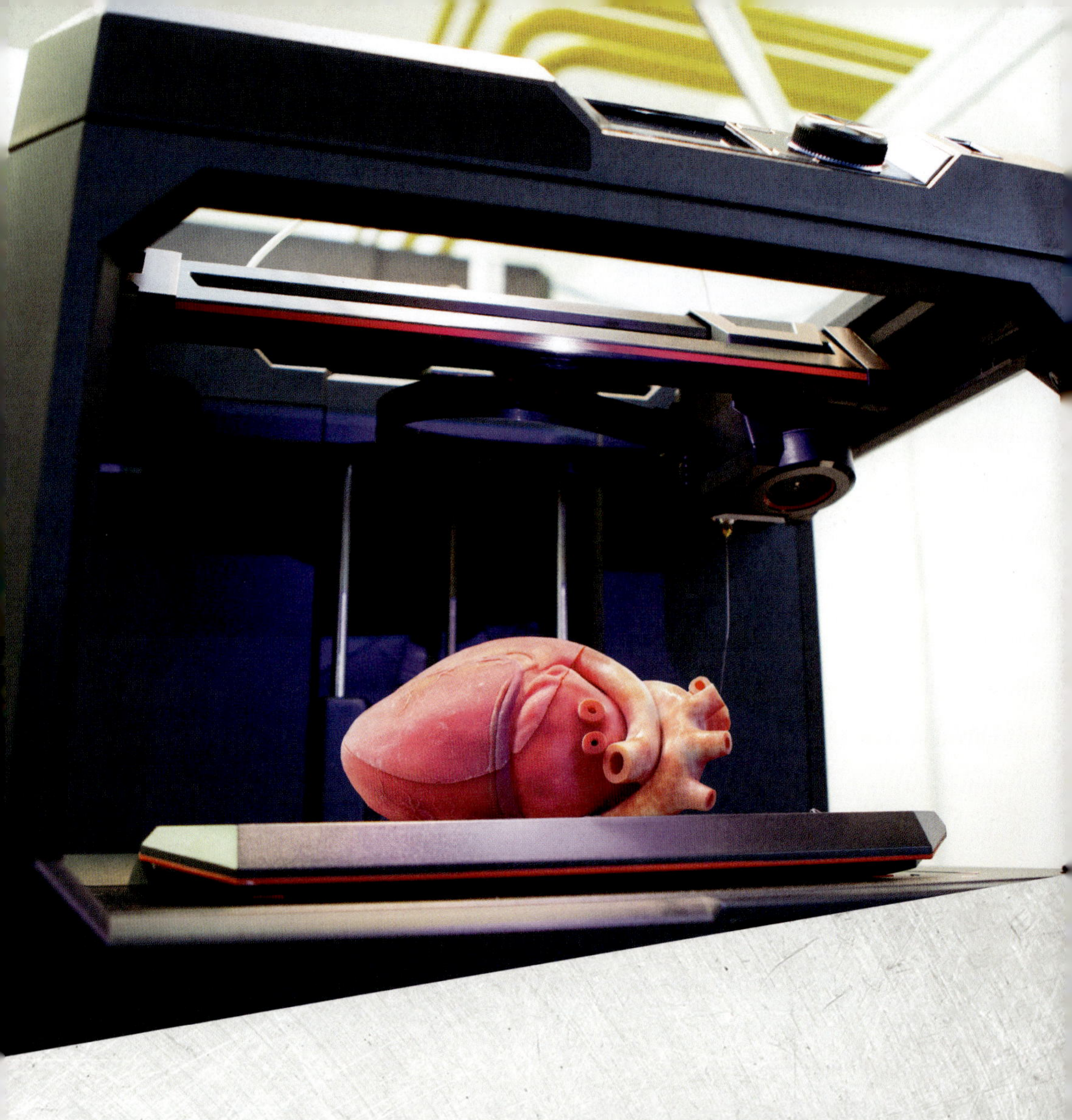

3D printers work by placing multiple layers of a material in order to create a 3D object. 3D printers can make dental implants. They can even make artificial hearts!

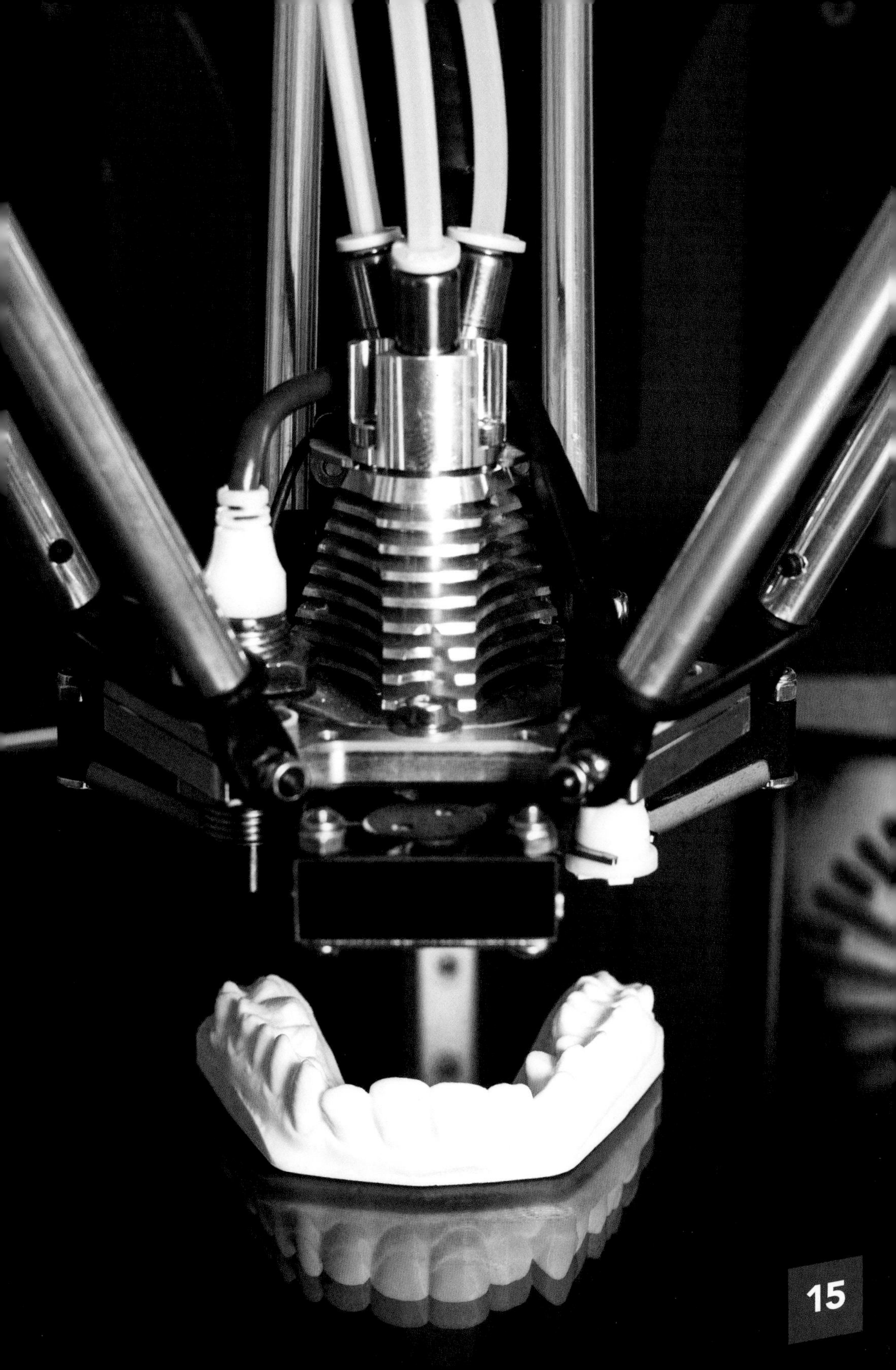

Robotic surgery is used every day. Small tools are attached to a robotic arm. A surgeon controls the arm with a **computer**. This allows for smaller cuts into the body. Patients can recover faster!

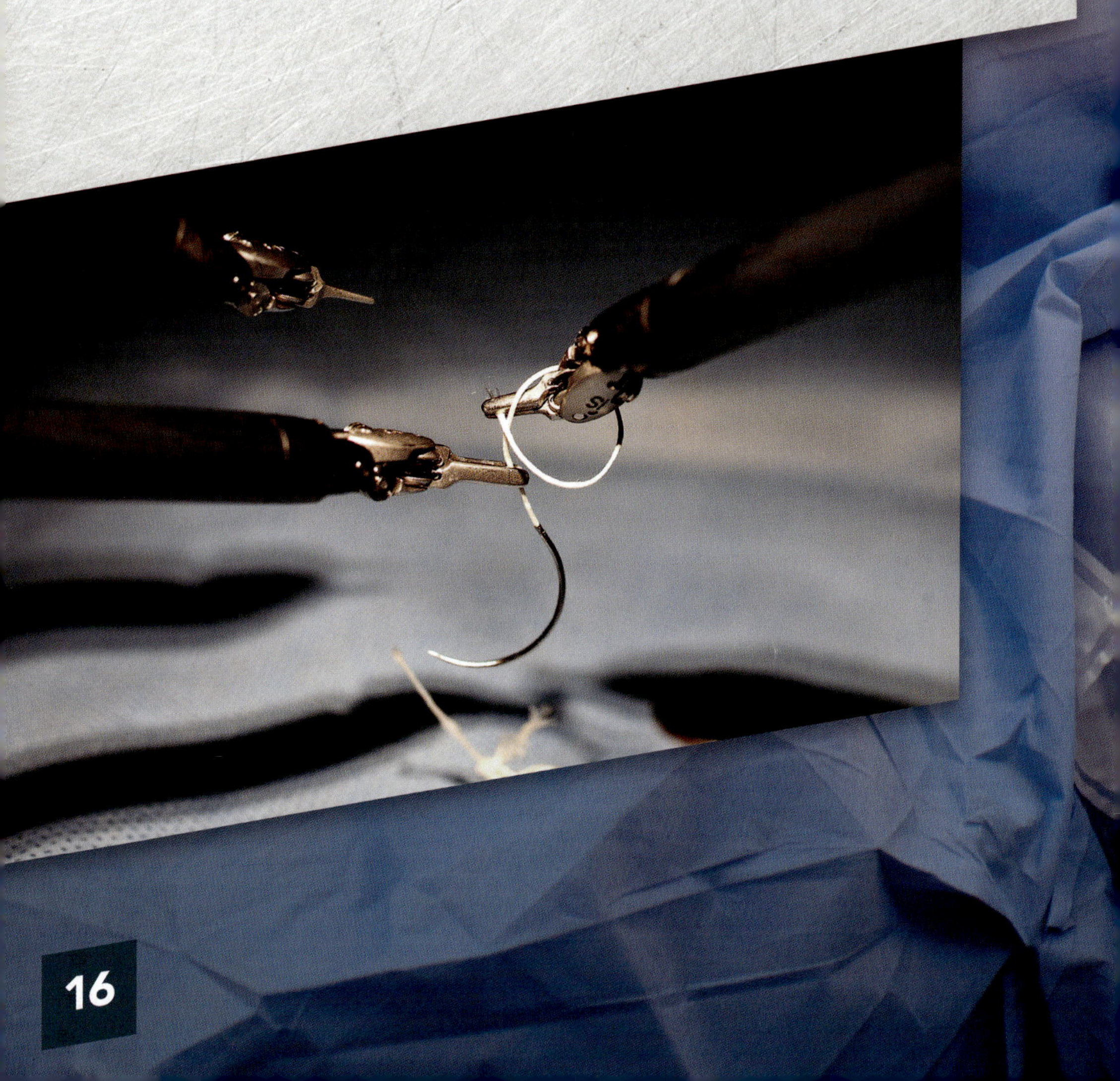

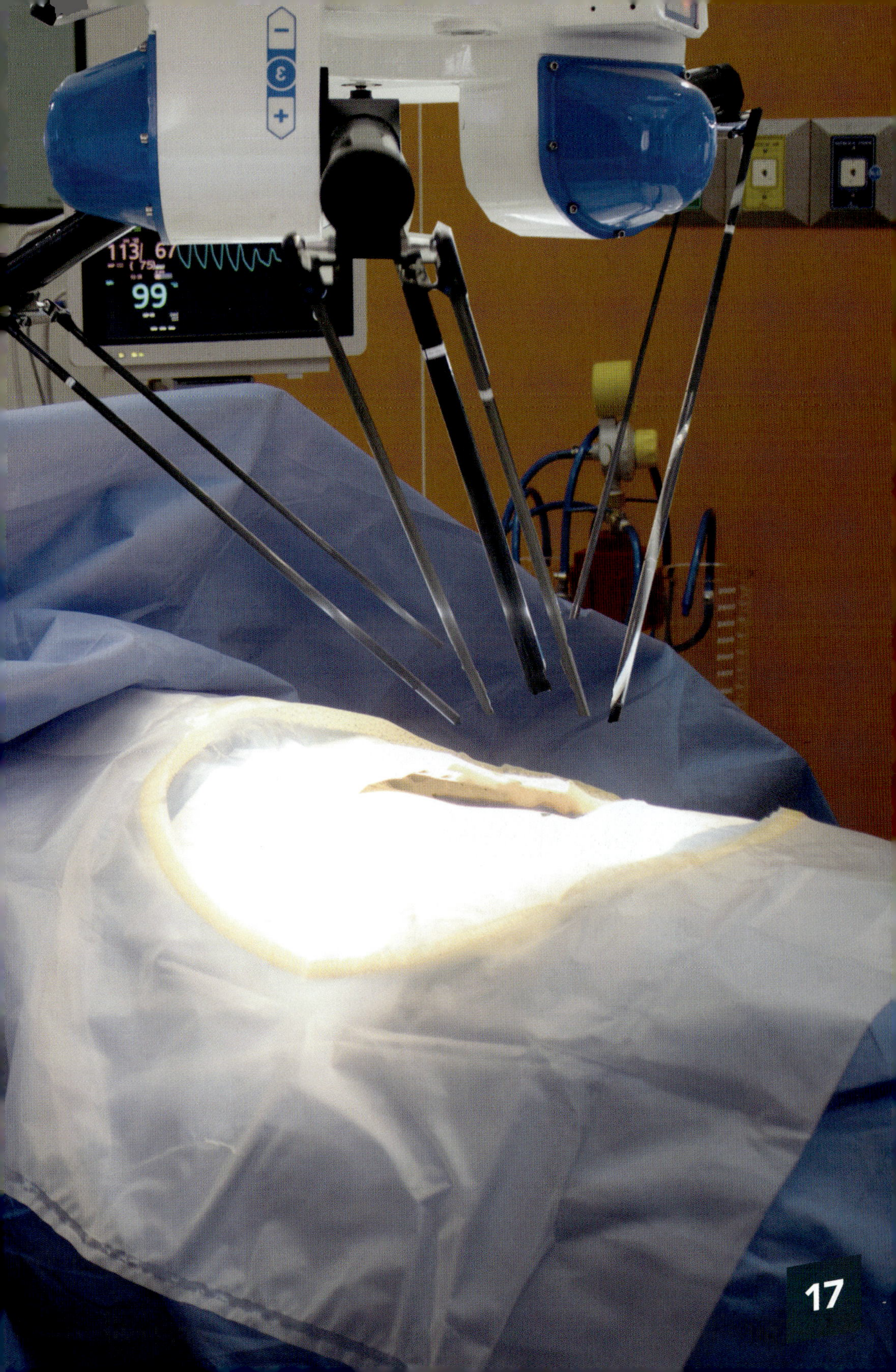
113 67
99

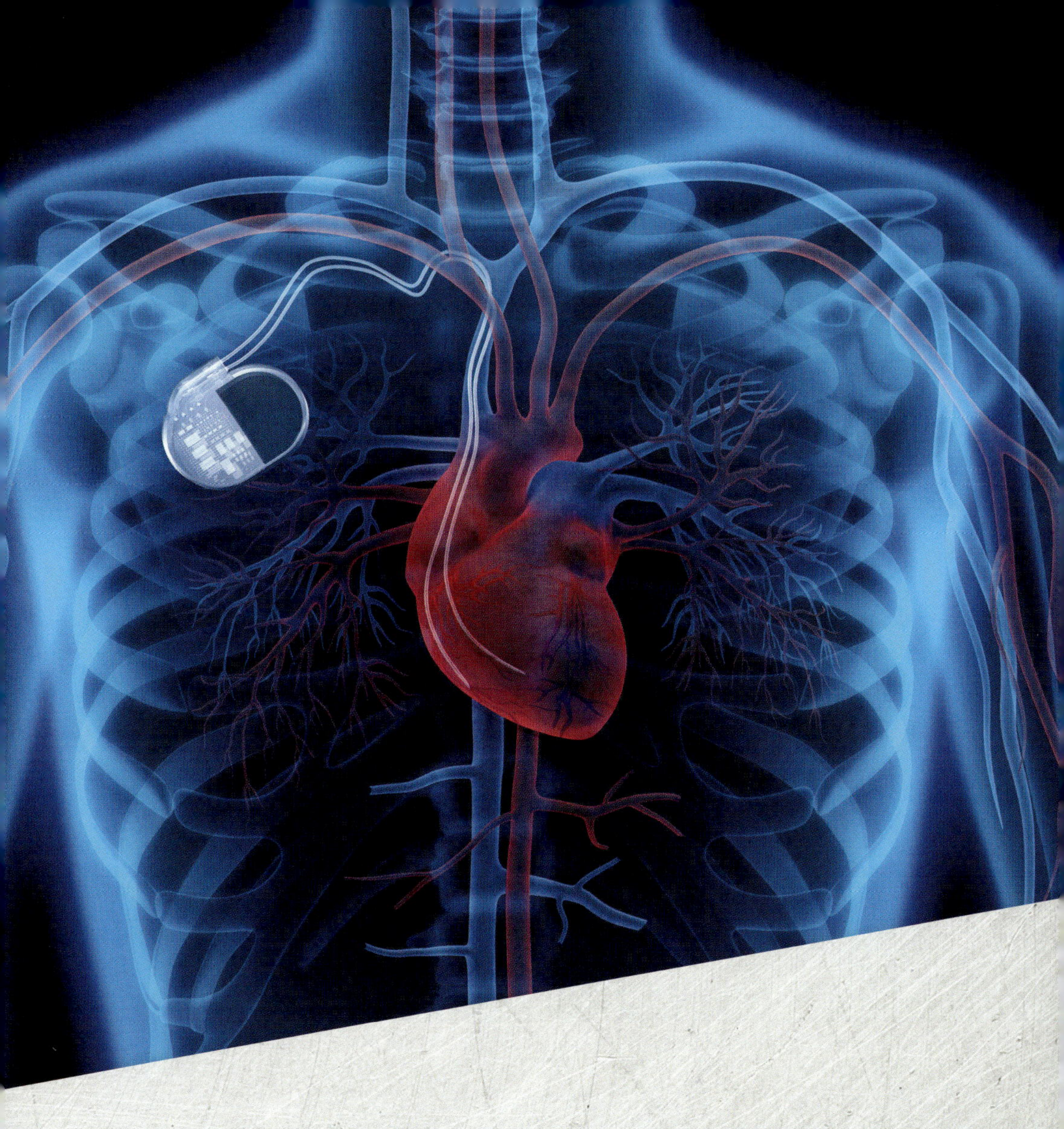

Artificial devices help many medical issues. They can replace damaged knees or hips. Cochlear implants help people hear. A pacemaker keeps a person's heart rate regular.

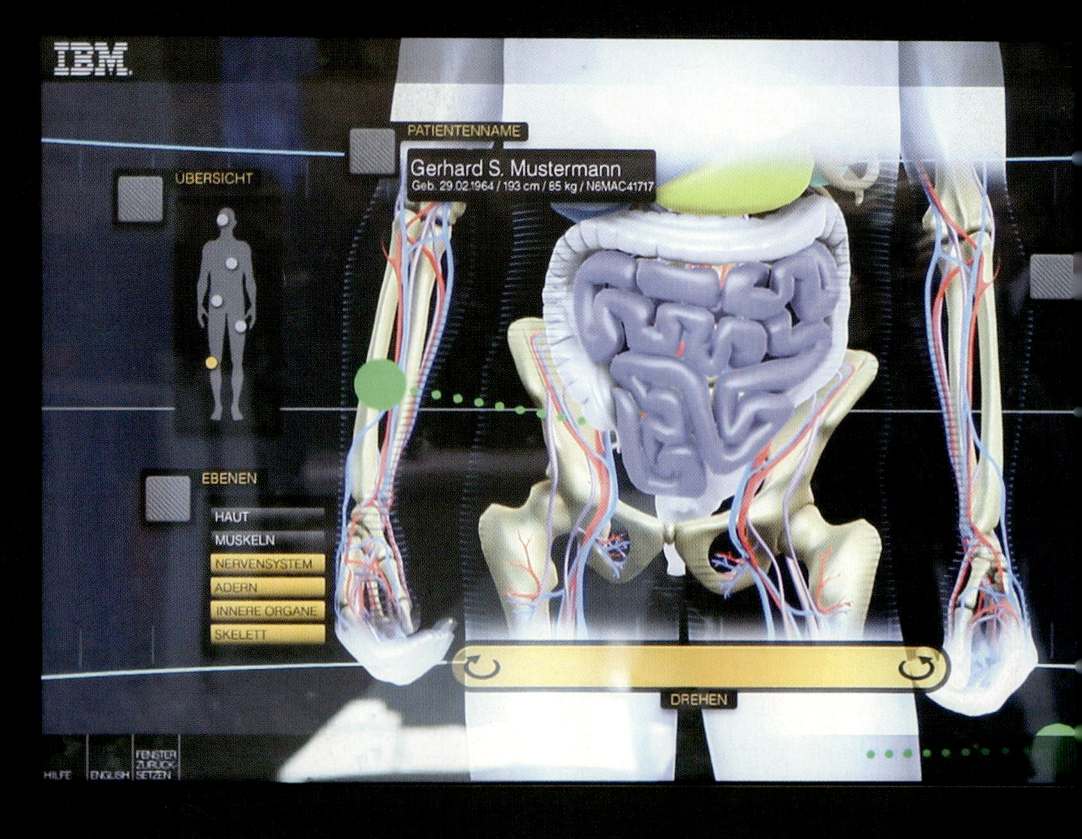
IBM
PATIENTENNAME
Gerhard S. Mustermann
Geb. 29.02.1964 / 193 cm / 85 kg / N6MAC41717
ÜBERSICHT
EBENEN
HAUT
MUSKELN
NERVENSYSTEM
ADERN
INNERE ORGANE
SKELETT
DREHEN
HILFE
ENGLISH
FENSTER ZURÜCK SETZEN

Many medical technologies are available. And more are being invented each day. They save and improve millions of lives around the world.

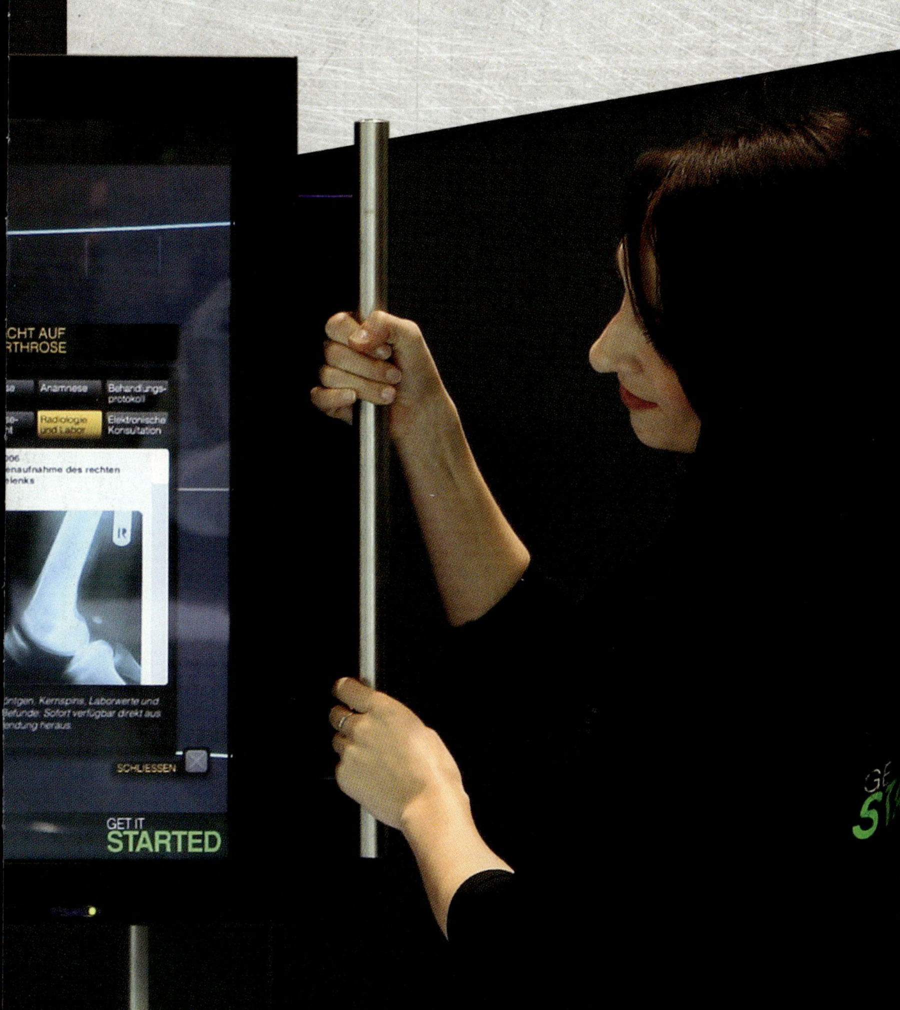

More Facts

- The first partial face transplant on a human was performed in 2005. The patient's face had been damaged from a dog attack.

- Penicillin was discovered in 1928. It was one of the first antibiotics. It treats infections that were once deadly.

- 3D printers can make custom pills that contain multiple medicines. This means people can take fewer pills.

Glossary

cancer – a disease in which certain cells divide and grow much faster than they normally do.

computer – an electronic device that is used to store and sort information and work with data at a high speed.

diagnose – to determine the identity of a disease by examination.

stethoscope – an instrument that makes the sounds inside a body louder. Doctors and nurses listen to the heart with a stethoscope.

x-ray – a beam of high energy radiation that is able to pass through many kinds of solid material.

Index

Online Resources

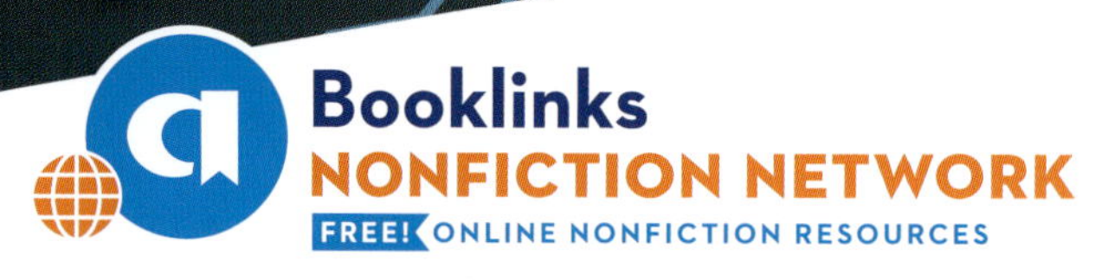

To learn more about medical technology, please visit **abdobooklinks.com** or scan this QR code. These links are routinely monitored and updated to provide the most current information available.